B&T
YOUNG ADULTS
STO

ALLEN COUNTY PUBLIC LIBRARY
P9-AFY-656

Greenhaven World History Progran

GENERAL EDITORS

Malcolm Yapp
Margaret Killingray
Edmund O'Connor

Cover design by John Castle

ISBN 0-89908-107-X Paper Edition
ISBN 0-89908-132-0 Library Edition

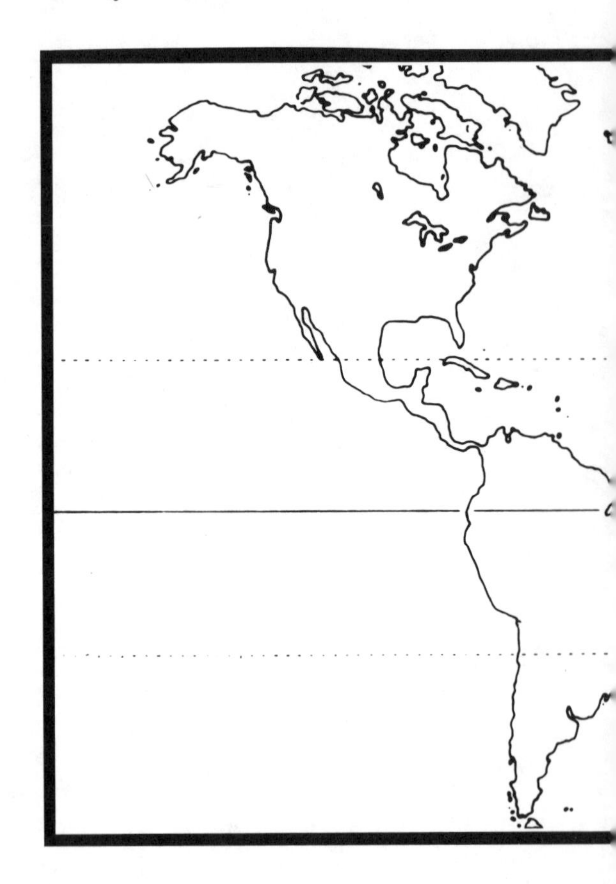

First published in Gratin Britain 1974 by
GEORGE G. HARRAP & CO. LTD

SCIENTIFIC REVOLUTION

by Peter Amey

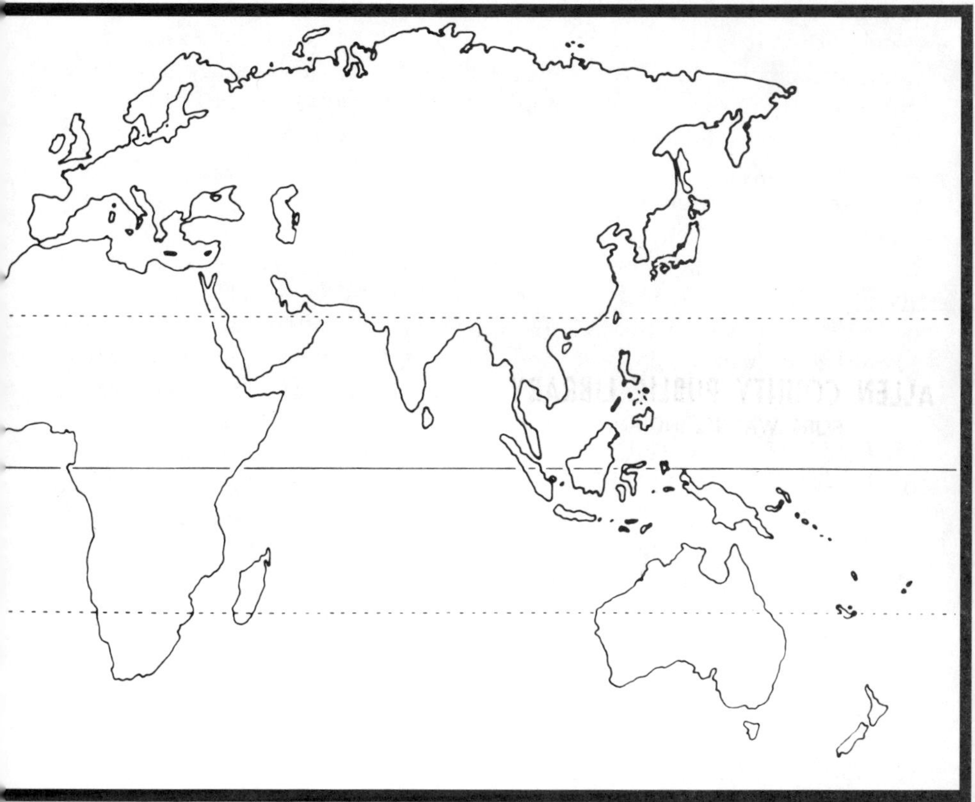

We know that the way we live is very different from the way people lived five hundred years ago. Our homes, our schools, where people work, the jobs they do, the ways we travel, our hospitals, what we do in our spare time, all these things are different. Science, working through technology, has played a very large part in bringing about these changes. Science and technology are closely connected. Scientific knowledge is knowing how and why everything behaves in the way it does. Technology is 'know-how'. It is knowing how to make and change things for useful purposes. By finding out how and why things behave in certain ways science provides the knowledge which can be used to increase our technology. The scientific revolution which came about in the sixteenth and seventeenth centuries, has made it possible for science to do this.

SCIENCE BEFORE THE SCIENTIFIC REVOLUTION

The study of science did not begin with the scientific revolution in Europe, because men in all ages and in all places have wanted scientific knowledge. Every great civilization has developed some science. For example, long before Christopher Columbus discovered the Americas, the Aztecs had developed skill in astronomy and mathematics, and, like the Incas *(Ancient America)**, knew how to make copper and bronze. Unfortunately, the Spaniards who invaded their lands were more impressed by the gold the Aztecs possessed, and by the human sacrifices they made to their sun-god. They killed many of the Aztecs, took their gold, and ignored their scientific knowledge until it was too late. When the Spaniards did finally come to take an interest in Aztec and Inca science, most of it had been forgotten. It could not help the scientific revolution.

THE INFLUENCE OF CHINA

One of the advantages that the Spanish had over the Aztecs was that the Spanish had guns. Gunpowder was developed in China. So were printing and the magnetic compass. The Arabs learned of all three things from the Chinese. The Europeans learned of them from the Arabs. All were well known in Europe by 1500. They are important in the history of science, as Francis Bacon pointed out long ago. (D1)* * Printing was important because, by making books in thousands, instead of copying them by hand, one by one, as they did before, knowledge could be spread faster to many more people.

Gunpowder and the magnetic compass were important in a different way. In a world of many wars, growing trade, and more travel, both were very useful. They were also puzzling. Why did a certain mixture of chemicals explode? What was the path of a cannon-ball through the air? What was the mysterious force which brought the compass needle round to the north? More scientific knowledge was needed to answer such questions.

*Titles in brackets refer to other booklets in the Program

Mons Meg: a fifteenth-century cannon

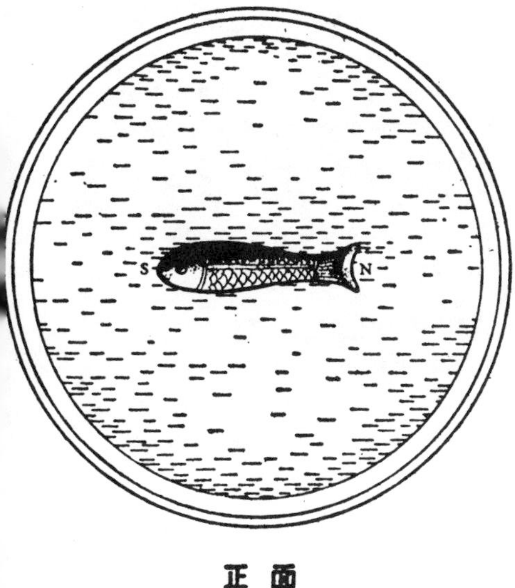

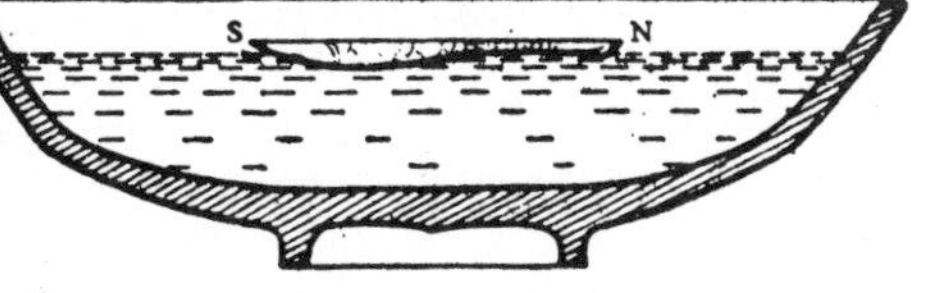

An early Chinese compass, 1044 A.D.

Discoveries such as that of gunpowder led to far fewer changes in China than in Europe. Traditional China was far more stable. In other words it was able to make use of new things without being upset by them. In Europe the same inventions led to the end of the armoured knight and his towering castle. It also led to many other changes which we shall mention later.

THE INFLUENCE OF INDIA

India is not famous for its science, yet it was India that gave the world its modern system of numbers. The clumsy Roman system has no sign for zero, and is very hard to work with (what is XVIII divided by IX?). The Hindu system of India has the sign '0' for zero, and has the simple numbers we still use today. The Arabs passed the Hindu system on to Europe and it gradually replaced the Roman

**The reference (D) indicates the numbered documents at the end of this book.

system from the twelfth century onwards. The new system helped scientists work out more difficult problems than they could before.

THE INFLUENCE OF THE GREEKS

The great age of Greek science began in the sixth century B.C. and continued for at least three hundred years. Although some Greek learning was lost for ever, and some was lost sight of by European scholars for several hundred years, much was written down and remembered. Among the most famous of the ancient Greeks were Plato and his pupil Aristotle. Both men were *philosophers* as much as they were scientists. In other words they wanted to know how and why things worked, but they were also interested in methods or ways of reasoning.

Many Greek thinkers were interested in mathematics, especially geometry. Plato was sure of its importance. Over the door of his academy or school was written 'Let no one enter here who has not learnt mathematics'. Plato believed that what our senses (such as our sight or hearing) tell us is not to be trusted. (D2) We can understand this. The moon looks bigger than the stars but we know it is much smaller. He also thought that physical work, work a man does with his muscles, spoils a man's soul. (D3) A man must have as perfect a mind and soul as possible if he is to understand the world, so a scientist must not do practical work. Plato therefore believed that experiments were not to be trusted, because our senses are not to be trusted. He also believed that experiments, which always need some practical work to carry them out, were not fit for a free born Greek to do. He taught that a scientist must rely mainly on pure *reason. (The Enlightenment)*

Aristotle was less interested in mathematics than Plato, but if anything was more interested in methods of reasoning. He believed in *observation,* that is, in looking carefully at things in order to learn about them. (D4) Like Plato, however, he thought that practical work could do more harm than good in reaching true understanding.

The reasoning of Plato and Aristotle was clear, and their explanation of the universe was complete and convincing. For hundreds of years men looked at the subject of science in the way they had taught.

THE COLLAPSE OF GREECE AND ROME

The Greek states were conquered by the Roman empire in the end, but for a time Greek learning was remembered and even added to. In turn, under pressure from the barbarians of the north and east, and from the Arabs in the south and east, the Roman empire collapsed. It took a long time, from about the fifth to the seventh centuries A.D.

Because of the collapse law and order could no longer be relied upon. The big cities of the empire shrank in size and importance. Some cities were abandoned altogether. Trade declined partly because robbers, pirates and warlords made it unsafe and unprofitable. Because trade declined men had to rely on producing things for themselves, or go without altogether. Even when Greek civilization had been at its peak only a few men in any time could specialize in learning. Now, by the seventh century A.D., there were fewer still. Even those who could specialize were often cut off from each other and from the books containing the learning of the past, by the difficulties of travel. Much of the old learning was lost, some for a time, and some for ever.

THE CHURCH

The Christian Church managed to keep some learning alive through the labours of its scholars. Most of these lived in monasteries. They spent part of each day in study, and in the slow, laborious work of copying books by hand. As Christians they believed that God had revealed the truth about His universe through the Bible. *(Religion)* **Christians believe** there is only one God, and as Plato and Aristotle had taught the same, it was not difficult for the Christian scholars to accept their teaching. For example, Aristotle had taught that the earth was at the centre of the universe. He said that the sun, moon, planets, and stars all revolved round the earth. This fitted nicely into the Christian belief that man was the most important of all God's creations. You would expect God to place man at the centre of everything!

The difficulties in keeping learning alive, the respect for Greek ideas, and the belief that God had revealed His truth in the Bible, together had an important consequence. They led men to think that all important knowledge was already known. Anything new would have to be fitted into what was already believed. As they had the wrong idea about many things this made progress difficult.

MEDIEVAL SOCIETY

Society in the middle ages (the medieval period) changed very slowly. Most people worked on the land in the same way as their parents and grandparents had done before them. They did not produce very much. Out of the little they had to spare they paid heavy taxes to their lord and to the Church. Kings and lords were usually more interested in ruling, enjoying themselves, and defending their power than in new ideas. They and the churchmen were the people who had spare money and time which they could have used for science, but apart from a few scholars, (D5) they were not interested in doing so. Science did not seem to be useful, and they did not have the imagination to see that one day it might be. So hardly any effort was made to increase scientific knowledge.

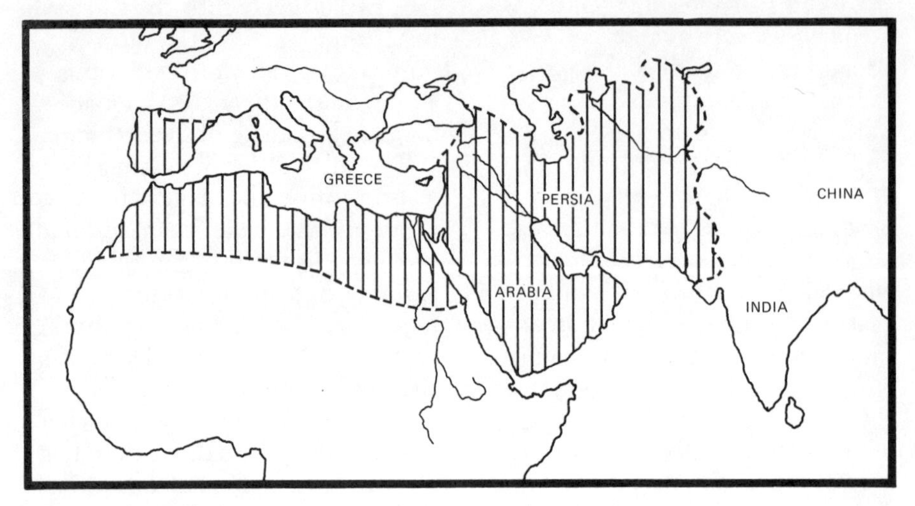

The Arab Empire, 732 A.D.

THE ARABS

Fortunately, at the very time when scientific learning was only just surviving in Christian Europe, Muslim learning was at its height. By 732 A.D., one hundred years after the death of their great prophet Muhammad, the Arabs had conquered a great empire for their religion of Islam. *(Muhammad and the Arab Empire)* It stretched from India in the east to Spain in the west. Their new empire contained many different peoples, including Indians, Persians, Greek Christians, and Jews. They traded with the Chinese in the east and the Europeans in the west. From about 750 A.D. to 1200 A.D. learning flourished in great Islamic centres such as Baghdad and Toledo, as scholars studied and translated Greek, Persian, and Indian writings.

Islamic scholars extended as well as preserved learning, especially mathematics, astronomy, medicine, and chemistry. They invented a number of scientific instruments. One of these was the *astrolabe.* This could be used to measure the angle of the stars above the horizon and for other astronomical calculations. The many Arab words used in science show how important the Arabs were. Chemical words such as alkali, mathematical words such as algebra, astronomical words such as zenith, all come from Arabic.

By preserving and handing on the ideas and discoveries of the Greeks, Indians, Persians, and Chinese, and by adding many of their own, the Arabs made a big contribution to the growth of science in the world.

LEARNING BEGINS TO GROW AGAIN IN EUROPE

From about 1100 onwards more and more Arab writings were translated into Latin by European scholars. Latin was the language of learning all over Europe in the middle ages. These translations helped Europeans to think for themselves. This was because ideas learnt from the Arabs were often different from the older ideas in Europe. Scholars were therefore forced to reason out which were right. It took a long time – about four hundred years from the twelfth century to the sixteenth century – before scholars and scientists could get used to thinking for themselves. It was so difficult to break away from the habit of believing that every important thing was already known.

Admiration for the ancient

Copy of an Arabian astrolabe, 1067 A.D. *The moveable hands measure the angles of the stars.*

Greeks increased in the late fourteenth century. Many works of Greek science, art, and literature were rediscovered. Scientists, artists, and writers copied Greek ways and produced great works of their own. To many it was if they were taking part in the re-birth or

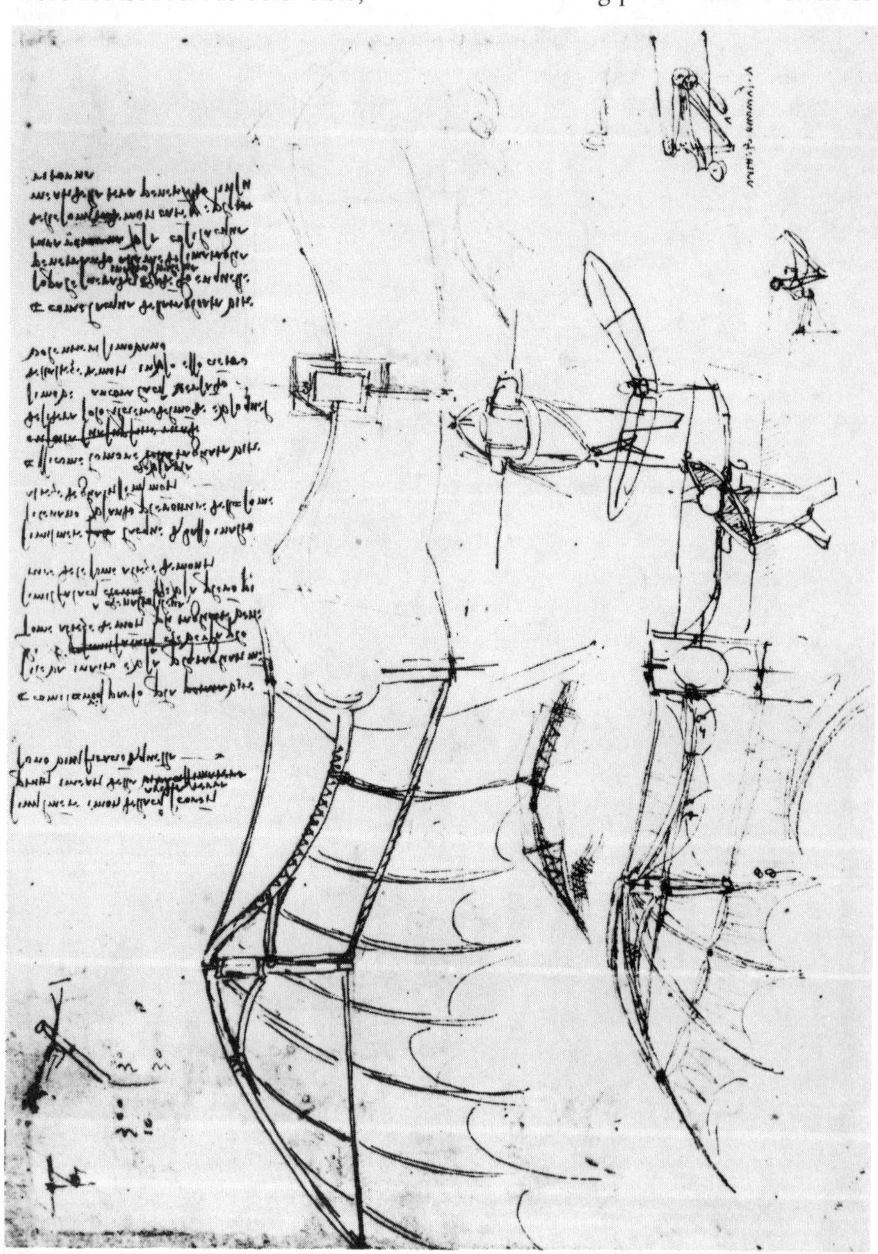

Design for a fixed-wing aircraft from the notebooks of Leonardo da Vinci. Leonardo's notebooks contain hundreds of drawings and notes on the many subjects that interested him.

renaissance of the splendid days of old.

One of the most famous men of the renaissance was Leonardo da Vinci who was as interested in science as he was in art. He studied all sorts of sciences such as anatomy, astronomy, mathematics, mechanics, and optics. He designed a flying machine (which was never built), and he made one of the best studies of birds in flight that there has ever been.

Leonardo was at the turning point in the revival of learning. He realised that Plato was wrong. Pure reason was not enough. Experiments were necessary to check on the ideas that reason suggested. The Greeks and the medieval scholars did not realise the importance of experiments, so in this he was more like a modern scientist. But he was interested in too many things. He was not methodical enough. He never concentrated on one problem long enough to work it out properly. Because of this he did not do much to help the scientific revolution.

IMPROVEMENTS IN TECHNOLOGY

In the three hundred years before the scientific revolution, trade and industry began to grow rapidly. More and more technology was developed in the process. Miners dug deeper and needed pumps to clear the water. So pumps were improved. Better methods were worked out to separate metal from the rock it was found in. Seamen began to venture further, often going out of sight of land. They began to use compasses, maps, and navigational instruments to find their way. Clothmakers developed better dyes for their cloths. Glassmakers made better glass. Spectacles, telescopes, and microscopes became possible. All these changes, and others like them, helped science because people could learn from experience.

ALCHEMY AND ASTROLOGY

Help was also given by the false sciences of *alchemy* and *astrology*. Alchemists thought they could make gold. They were wrong, but they did invent *distilling*, which is a very useful way of separating liquids by heating them. Astrologers thought they could tell peoples' fortunes by the stars. They too were wrong, but they did add to our knowledge of the movements of the stars and planets.

THE SCIENTIFIC REVOLUTION BEGINS

Gradually scientists came to challenge more and more what the ancients taught. They came to develop new, better methods of finding out how things worked. Mathematical knowledge increased and helped them to reason. They began to think up experiments to check on their ideas in a methodical way. The scientific revolution had begun.

Many men were needed to bring this about. These men came from every part of Europe. They wrote books to explain their ideas. The printing press made it possible to produce thousands of copies which found their way all over Europe.

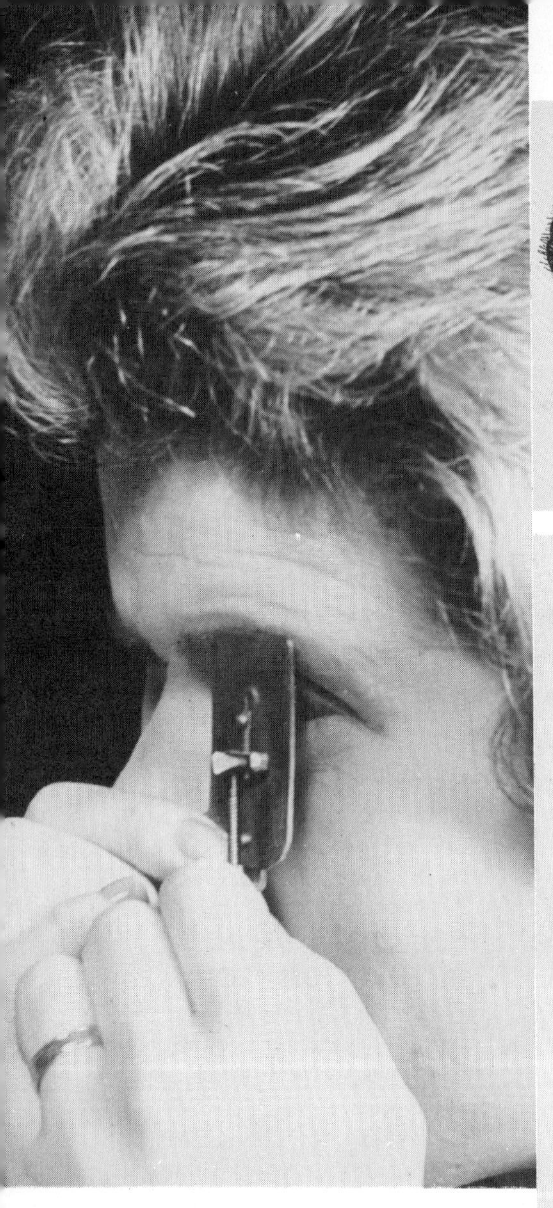

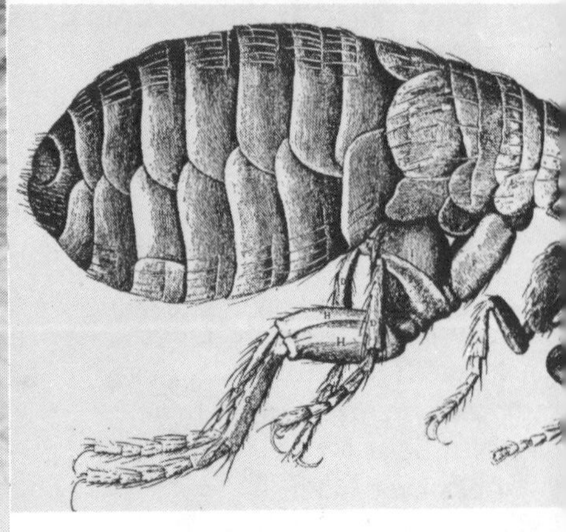

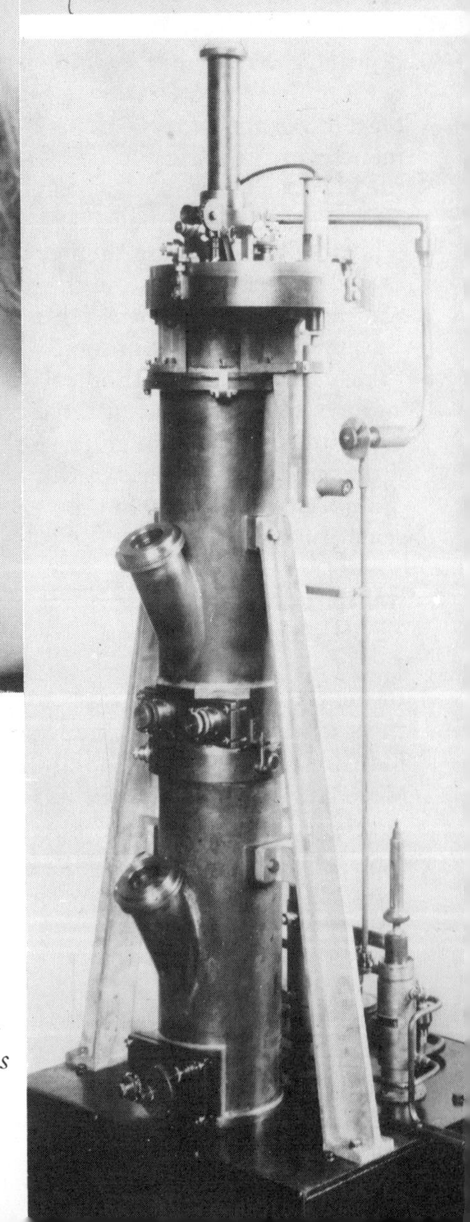

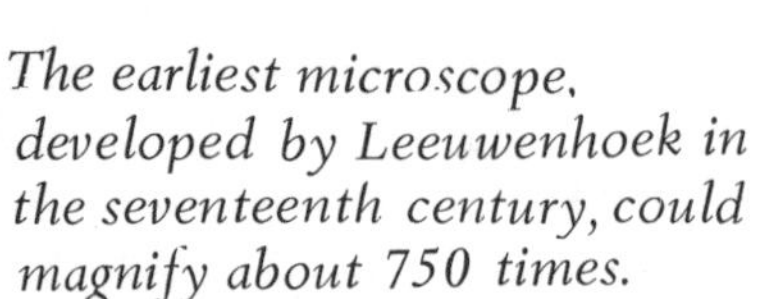

The earliest microscope, developed by Leeuwenhoek in the seventeenth century, could magnify about 750 times.

This is how a flea would have looked through an early microscope. (Upper right)

One of the first electron microscopes. Today these microscopes magnify over 3,000,000 times.

This picture of 1537 shows an alchemist and his assistant. Their ragged clothes and the crowded room show that they have not yet found the secret of making gold.

Scientists were able to learn from one another and give one another new ideas. So the scientific revolution was not the work of Englishmen, or Frenchmen, or Italians alone. It was the work of Europeans. And, as we have seen, even they did not do it all by themselves. The Chinese, the Indians, the Persians, and the Arabs all gave something before it came about. Today this is not hard to understand, because men and women from all over the world add to scientific knowledge and so help one another.

COPERNICUS, GALILEO, AND NEWTON

The scientific revolution began with the Pole, Nicolas Copernicus (1473-1543) and was carried on by many other men including the Italian Galileo Galilei (1564-1642) and the Englishman, Isaac Newton (1642-1727). Copernicus, Galileo, and Newton showed that Aristotle's ideas or *theories* of the universe were wrong. These three men were very important because each was bold enough to use his imagination to find a new theory. He then checked the theory as carefully as he could. Copernicus used observations and mathematics to do this. Galileo and Newton who came after him did the same, and they also used experiments. (D6)

Aristotle had taught that the earth was the centre of the universe. The sun, moon, planets and stars all moved round it. He said that nothing moved unless something was pushing or pulling it. It followed therefore that the sun, moon, planets and stars must be fixed in invisible spheres which

This fifteenth-century printing press is the type used by William Caxton, the first man to print books in England.

moved them round in the sky. Copernicus showed that it was simpler to think of the earth as a planet moving round the sun. Many people agreed that it was simpler but they did not accept that it actually happened. One of Galileo's great achievements was to show that it did. He showed this in several ways. For example, people said that the earth could not move round in space because if it did it would leave the moon behind. Galileo built one of the first telescopes and with it he discovered the moons of Jupiter. He pointed out that if Jupiter could move through space taking its moons with it, so could the earth.

Galileo also showed that Aristotle's ideas about movement were wrong. Aristotle had said

Nicolas Copernicus

Galileo Galilei

that anything moving needed some force to keep it moving or else it would stop. Galileo was able to show that a moving object goes on moving unless something else stops it. However he was wrong when he said that the moon, planets, and stars naturally move in a circle. Newton, by making use of Galileo's ideas and adding some of his own, gave a much better explanation.

He realised firstly that anything keeps still or keeps moving in a straight line until something happens to change it. But if this were so, why doesn't the moon, which is moving in space, travel in a straight line away from the earth? There is a story that he saw an apple falling from a tree and this gave him the idea he needed: the same gravity that pulled down the apple kept the moon from moving away. Newton then developed some advanced mathematics. With this he proved his theory. It was the pull of gravity which kept the moon from leaving the earth, and the planets from leaving the sun.

Sir Isaac Newton

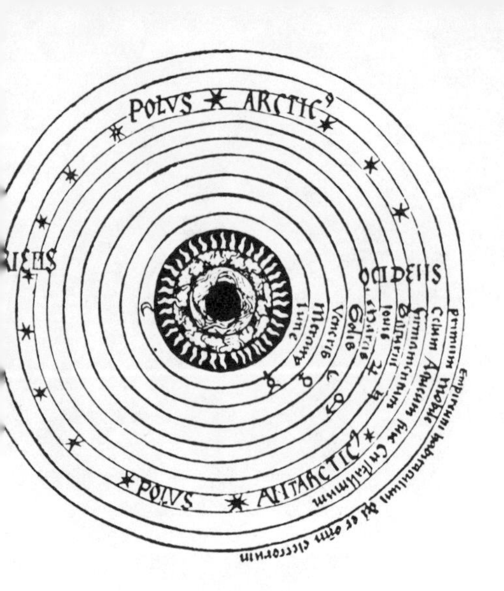

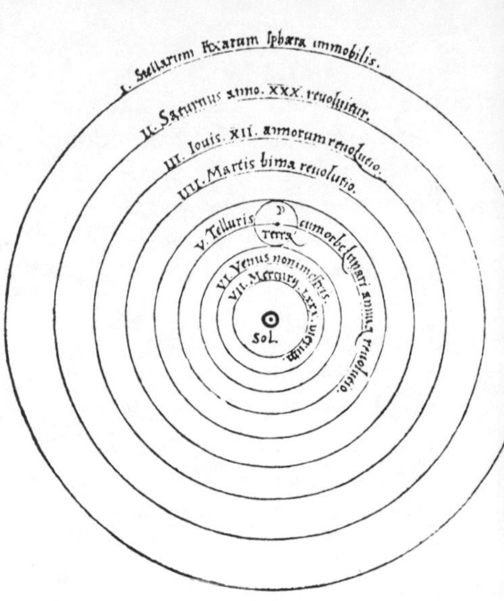

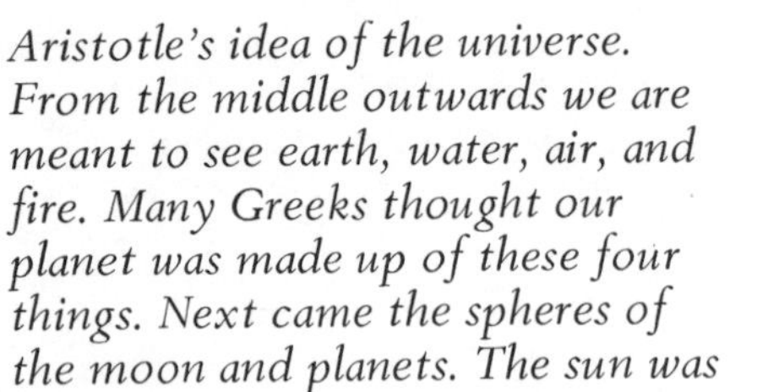

Aristotle's idea of the universe. From the middle outwards we are meant to see earth, water, air, and fire. Many Greeks thought our planet was made up of these four things. Next came the spheres of the moon and planets. The sun was shown between Venus and Mars. This drawing was made in 1504 A.D.

Copernicus' idea of the universe, 1543 A.D. *Sol, the sun, is at the centre. Terra; the earth, with the moon in orbit, is the third planet away from the sun.*

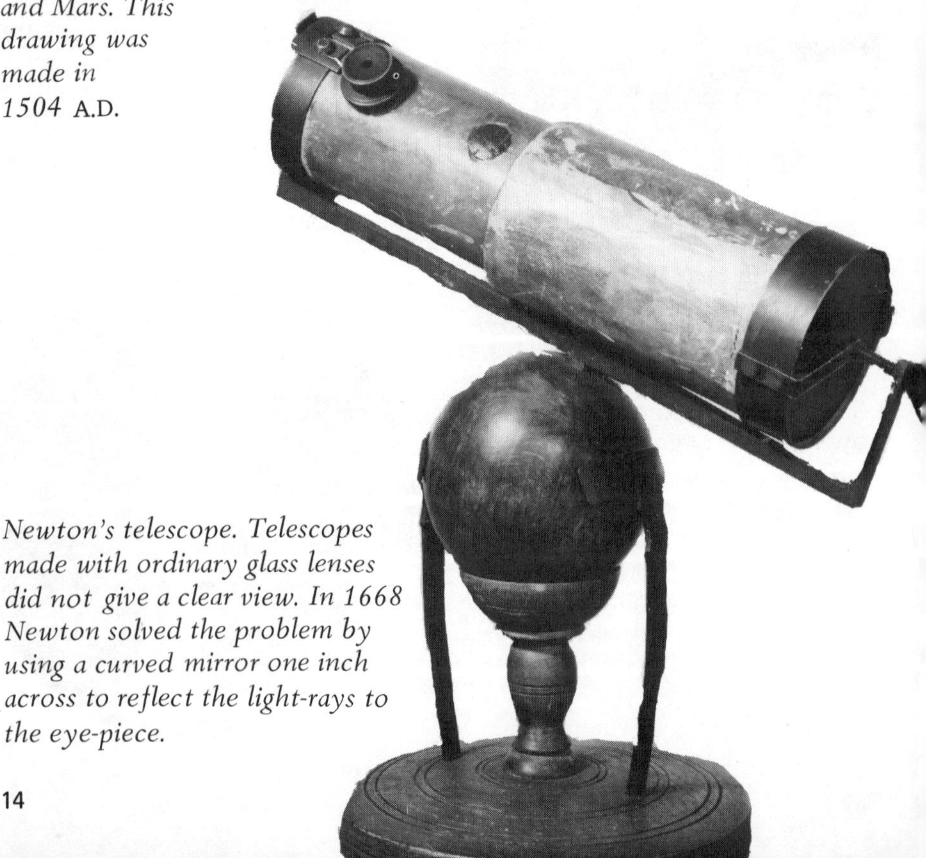

Newton's telescope. Telescopes made with ordinary glass lenses did not give a clear view. In 1668 Newton solved the problem by using a curved mirror one inch across to reflect the light-rays to the eye-piece.

William Herschel's giant forty-feet telescope, 1795. William Herschel, a German who lived in England, used Newton's idea to build a telescope with a mirror four feet across. The tube, forty feet long, was moved up and down by ropes and pulleys.

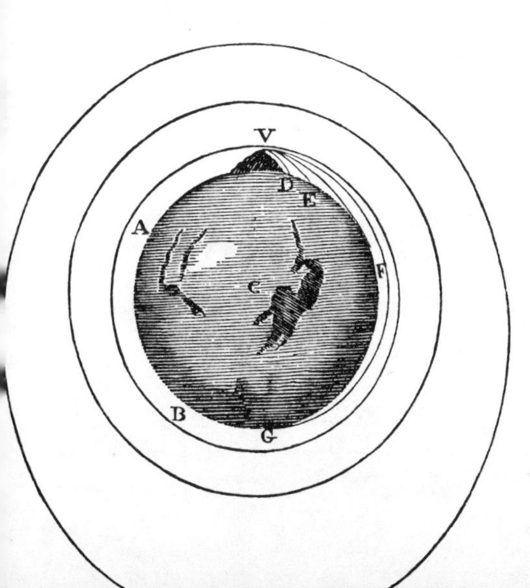

The influence of gravity. (This is the diagram to which Newton refers in document 7)

This work was important not just for astronomy. It had shown that the Greeks were wrong. People realised that if the Greeks were wrong in one science they might be wrong in others. So every science was studied in the new way.

Two other important men of the scientific revolution were the Englishman Francis Bacon (1561-1626) and the Frenchman René Descartes (1596-1650). Francis Bacon believed that the universe and everything in it worked according to certain rules or laws. These laws were sometimes called the laws of nature. They were the laws by which God ran the universe. He said that scientists should collect all the facts they could. They should then put these facts into order. This would suggest various theories. These theories should be tested by experiments. The results would lead to the laws being discovered. (D9)

Francis Bacon

René Descartes was also very interested in how scientists should reason things. (D10) It was obvious to his mind that the universe was like a great machine with millions of parts, put together and made to work by God. Reason, helped by mathematics and experiments, would discover the laws which God made everything obey.

Sir Isaac Newton thought the same way as Bacon and Descartes. The laws of gravity which he worked out seemed to explain perfectly the way the planets moved round the sun. He believed that they were God's laws. They made him very famous. Alexander Pope, a famous poet of Newton's time wrote:

> Nature and Nature's laws lay hid in night:
> God said, *Let Newton Be!* and all was light.

The light that Newton, Descartes, and other scientists brought was the light of science and reason. Encouraged by Newton's success men began to think that science and reason could find out the answer to everything. They would end the darkness of ignorance and enlighten the world. *(The Enlightenment)*

SCIENTIFIC SOCIETIES

In the seventeenth century, the century of Galileo and Newton, there were still only a very few scientists. Because travel was difficult it was not easy for them to meet and learn from one another. Some science was studied at the universities. Galileo studied at Padua University and Newton at

René Descartes

Cambridge. But most people at the universities were more interested in other things such as religion.

Science needed more help to grow. In 1660 some English scientists had the idea of forming a society to help science. It was called the Royal Society because King Charles II was a member. (D11) In 1666 the Academy of Sciences was formed in Paris. In the next one hundred years many other countries in Europe started scientific societies. These societies helped science in two important ways. Firstly they collected facts and conducted experiments. Secondly they published journals (something like magazines) so that other scientists could learn the latest ideas and discoveries from them. (D12)

PATRONAGE

The Royal Society and the Academy of Sciences were fortunate because they enjoyed royal *patronage.* Patronage means giving help like a father gives help to his child. Royal patronage was important because in the seventeenth century kings governed their countries and so had a lot of power. Governments have given help to science ever since. Nowadays they provide most of the money out of taxes for the scientists to carry on their work. They spend this money because scientific knowledge is often useful. For example Charles II set up the Royal Observatory at Greenwich to make exact lists of the positions of the stars, planets and moon. These lists were needed by seamen to help them *navigate,* that is, to find their way.

Today governments all over the world spend money on finding out scientific knowledge. This is called scientific research. Some research is so expensive that only governments of rich countries can afford it. For example space research and research into the secrets of the atom are both very expensive. Other scientific research is carried

A giant burning glass of the eighteenth century. This glass belonged to the French Academy of Sciences. The glass was mounted on wheels so that it could be kept facing the sun, and was one way of obtaining the great heat needed for some experiments.

out by commercial companies. These are companies that do things for a profit. If they can discover something new that people want such companies can often make a lot of money.

THE DEVELOPMENT OF DIFFERENT SCIENCES

Since the scientific revolution many different sciences have grown up using its methods. In the age of the enlightenment men such as Adam Smith and Thomas Malthus even began to use the new scientific methods in the study of economics. This is the study of how mankind makes its living. We now think of economics as one of the 'social sciences' which are concerned with how society, that is all of us together, behaves. At first people hoped that economics could be studied like any other science. But nowadays we believe that, like the other social sciences, it cannot be so exact because people vary so much in the way they behave.

We call sciences like chemistry, biology, and physics, 'natural sciences' because they are to do with things found in nature, from the stars downwards. The scientific revolution has led to great advances in the natural sciences.

Chemists study the ways chemicals go together to make up all the different substances we know of. The Frenchman, Antoine Lavoisier (1743-1794), is thought of as the father of modern chemistry because he realised that some substances are formed of only one chemical. These are called elements. Other substances are made up of two or more elements. Lavoisier saw how useful it would be for the understanding of chemistry if all the elements were sorted out and listed. He found about twenty. Scientists now know of about one hundred.

Biology is another science which has made progress since the scientific revolution. Biology is the study of living things like plants, insects, and animals. Charles Darwin, who lived in the nineteenth century, did a lot to develop this science. He sailed round the world in a ship called the *Beagle* and collected a huge amount of facts about the animals and plants he saw. (D13) Perhaps he is best remembered for his suggestion that human beings developed from a sort of man-like animal many hundreds of thousands of years ago. This suggestion caused a tremendous fuss. Most European people of his day still thought that God made man in one day as it says in the Bible. But Darwin, and many scientists since, have found a lot of facts which show that he was right.

The most famous woman scientist is the Polish woman Marie Curie (1867-1934) who married a Frenchman. She discovered radium, a very dangerous radio-active substance. Although it is dangerous it is put to the very good use of fighting cancer.

Another scientist who lived about the same time as Madame Curie was the Swiss Albert Einstein (1879-1955). He was a great mathematician and physicist.

Physicists are scientists who study matter and energy. Older theories of physics had explained quite well how the universe that people saw around them, worked. But they did not explain the new observations of distant stars and tiny particles. Einstein, with his new theories, helped to explain these and to lay the foundations for all the recent developments of atomic physics.

This cartoon of 1861 shows the public interest that Darwin's ideas on evolution had aroused.

THE INFLUENCE OF SCIENCE

The scientific revolution has made a lot of difference to many peoples' lives. From the eighteenth century science has helped industry to grow by finding out better ways of making things. *(The Industrial Revolution)* Because of scientific research we now have many kinds of useful machines, metals and chemicals. Scientists have found out how to use steam power to make electricity. They have shown us how to find oil and to turn it into petrol and plastics. Knowledge built up since the scientific revolution has led to the invention of the telephone, radio, and television. Science has shown us how to fly, and how to dive safely into the depths of the sea.

Scientific research has also brought about very big changes in farming. *(The Agricultural Revolution)* Scientists have found out the chemicals plants like to feed upon. Millions of tons of these chemicals, called fertilizers, are now made to help the farmer to grow more food. Different chemicals, called pesticides, can be sprayed on crops to kill harmful insects. Science has also shown how to breed better animals and plants, that grow more quickly and do not catch diseases easily.

Science has found the way to cure many illnesses. Scientists have discovered that dirt spreads disease. They have shown us the rules of good health. Those of us who live in rich countries like Britain can hope to live twice as long as people usually lived two hundred years ago. *(Health and Wealth)*

Many people have stopped believing in God because of science. They say that science shows that everything came about by chance. Others still believe in God. They say that science shows us how God made the universe, and what things He has made possible.

All these changes and many others began in Europe, because that is where the scientific revolution started. Many of these changes have since happened in other parts of the world as well. One reason for this is because Europeans travelled abroad and took their knowledge with them. Another reason is that science helps to make a country better off, so that every country tries to have some scientists of its own.

Much of what science has brought about is good. It is good that we can live longer, have enough food, and have plenty of things to enjoy. But science has also given mankind fresh problems. It has led to a great increase in the world's population which is still going on. Medical care keeps people alive, but many people in poor countries do not get enough to eat. Many people are crowded into cities in order to help make all the things we use. Life is sometimes not very pleasant for them.

There is a danger that we shall use up all the useful materials in the world, like oil, and copper, and trees, in making all the things we want. We poison the air with the gases that our car engines give off. We poison the waters of the world with the waste from our factories. This is often called pollution. Many animals and plants are in danger of being killed off by the damage we do. We might even kill everyone with the nuclear bombs that science has shown us how to make.

Science has found out the answer to many of our problems. It may find the answers to all of them, or it may not. Whatever happens, we can see that the scientific revolution is still playing a big part in our lives. (D14)

DOCUMENT 1

PRINTING, GUNPOWDER AND THE MAGNET

FRANCIS BACON (1561-1626) – Lawyer who became Lord Chancellor to King James I of England. He did not know the origin of these discoveries, but he realised their importance.

It is well to observe the force and virtue and consequence of discoveries. These are to be seen nowhere more clearly than in those three which were unknown to the ancients (the Greeks), and of which the origin, though recent, is obscure and inglorious; namely, printing, gunpowder, and the magnet. For these three have changed the whole face and state of things throughout the world, the first in literature, the second in warfare, the third in navigation; whence have followed innumerable changes; insomuch that no empire, no sect, no star, seems to have exerted greater power and influence in human affairs than these three mechanical discoveries.

DOCUMENT 2

THE SENSES

PLATO – A Greek philosopher of the fourth century B.C., here using a story or allegory to explain how our senses cannot be trusted to give us a true understanding of how things really are

Compare our natural condition . . . to a state of things like the following. Imagine a number of men living in an underground cavernous chamber, with an entrance open to the light, extending along the entire length of the cavern, in which they have been confined, from their childhood, with their legs and necks so shackled, that they are obliged to sit still and look straight forwards, because their chains render it impossible for them to turn their heads round: and imagine a bright fire burning some way off, above and behind them, and an elevated roadway passing between the fire and the prisoners, with a low wall built along it, like the screens which conjurors put up in front of their audience, and above which they exhibit their wonders.

I have it, he replied.

Also imagine a number of persons walking behind the wall, and carrying with them statues of men, and images of other animals . . . together with various other articles, which overtop the wall; and, as you might expect, let some of the passers-by be talking, and others silent.

You are describing a strange scene, and strange prisoners.

They resemble us, I replied. For let me ask you, in the first place, whether persons so confined could have seen anything of themselves or of each other, beyond the shadows thrown by the fire upon the part of the cavern facing them?

Certainly not, if you suppose them to have been compelled all their lifetime to keep their heads unmoved.

And is not their knowledge of things carried past them equally limited?

Unquestionably it is.

And if they were able to converse with one another, do you not think they would be in the habit of giving names to the objects which they saw before them?

Doubtless they would.

Again: if their prison-house returned an echo from the part facing them, whenever one of the passers-by opened his lips, to what, let me ask you, could they refer the voice, if not to the shadow which was passing?

Unquestionably they would refer it to that.

Then surely such persons would hold the shadows . . . to be the only realities.

Without a doubt they would.

DOCUMENT 3

PHYSICAL WORK *XENOPHON (430-355 B.C.) – Greek soldier and historian, here showing that Plato was like the other Greeks of his day in his attitude to physical work*

What are called the mechanical arts, carry a social stigma and are rightly dishonoured in our cities. For these arts damage the bodies of those who work at them or who act as overseers, by compelling them to a sedentary life and to an indoor life, and, in some cases, to spend the whole day by the fire. This physical degeneration results also in deterioration of the soul. Furthermore, the workers at these trades simply have not got the time to perform the office of friendship or citizenship. Consequently they are looked on as bad friends and bad patriots, and in some cities, especially the warlike ones, it is not legal for a citizen to ply a mechanical trade.

DOCUMENT 4

HABITS OF CRUSTACEANS *ARISTOTLE – Greek philosopher (384-322 B.C.), showing us the care with which he made his observations*

Crustaceans feed in like manner. They are omnivorous; that is to say, they live on stones, slime, seaweed, and excrement [animal waste] – as for instance the rock crab – and are also carnivorous [flesh-eating]. The craw-fish or spiny lobster can get the better of fishes even of the larger species, though in some of them it occasionally finds more than its match. Thus this animal is so over-mastered and cowed by the octopus that it dies of terror if it becomes aware of an octopus in the same net with itself. The crawfish can master the conger-eel, for owing to the rough spines of the crawfish the eel cannot slip away, and elude its hold. The conger-eel, however, devours the octopus, for owing to the slipperiness of its antagonist the octopus can make nothing of it. The crawfish feeds on little fish, capturing them beside its hole or dwelling-place; for, by the way, it is found out at sea on rough and stony bottoms, and in such places it makes its den. Whatever it catches, it puts into its mouth with its pincer-like claws, like the common crab. Its nature is to walk straight forward when it has nothing to fear, with its feelers hanging sideways; if it be frightened, it makes its escape backwards, darting off to a great distance. These animals fight one another with their claws, just as rams fight with their horns, raising them and striking their opponents; they are often also seen crowded together in herds.

DOCUMENT 5

THE POSSIBILITIES OF SCIENCE *ROGER BACON (1214-1292)*

– A Franciscan friar who taught at the universities of Paris and Oxford, here foreseeing a number of modern scientific inventions.

Machines for navigation can be made without rowers so that the largest ships on rivers or seas will be moved by a single man in charge with greater velocity (speed) than if they were full of men. Also cars can be made so that without animals they will move with unbelievable rapidity . . . Also flying machines can be constructed so that a man sits in the midst of the machine revolving some engine by which artificial wings are made to beat the air like a flying bird. Also a machine small in size for raising or lowering enormous weights, than which nothing is more useful in emergencies. For by a machine three fingers high and wide a man could free himself and his friends from all danger of prison and rise and descend. Also a machine can easily be made by which one man can draw a thousand to himself by violence against their wills, and attract other things in like manner. Also machines can be made for walking in the sea and rivers, even to the bottom without danger. For Alexander the Great employed such, that he might see the secrets of the deep, as Ethicus the astronomer tells. These machines were made in antiquity [in the days of Greece and Rome] and they have certainly been made in our times, except possibly the flying machine which I have not seen nor do I know anyone who has, but I know an expert who has thought out the way to make one. And such things can be made almost without limit, for instance, bridges across rivers without piers or supports, and mechanisms, and unheard of engines.

DOCUMENT 6

SCIENTIFIC EXPERIMENTS *GALILEO GALILEI – Professor of Physics and Military Engineering at the University of Padua, he realised the importance of mathematics and of experiments in science.*

Philosophy [Galileo meant science] is written in mathematical language, and the letters are triangles, circles and other geometrical figures, without which means it is humanly impossible to comprehend a single word . . . I know very well that one sole experiment, or concludant demonstration, produced on the contrary part, suffices to batter to the ground . . . a thousand . . . probable arguments.

DOCUMENT 7

THE LAWS OF GRAVITY *NEWTON – In this excerpt from* Principia *(1687) he goes on to explain how the velocity (speed) of planets and the pull of gravity between them keep planets in orbit, so that they do not fly off into space or fall into the sun.*

Law I: Every body continues in its state of rest, or uniform motion in a right line [a straight line] unless it is compelled to change that state by forces impressed upon it. . . .
Law II: The change of motion is proportional to the motive force impressed; and is made in the direction of the right line in which that force is impressed. . . .
Law III: To every action there is always opposed an equal reaction . . . .
The action of centripetal forces [gravity].
That by means of centripetal forces the planets may be retained in certain orbits, we may easily understand, if we consider the motions or projectiles (things that are thrown); for a stone that is projected is by the pressure of its own weight forced out of the rectilinear (straight) path, which by the initial projection along it should have pursued, and made to describe a curved line in the air; and through that crooked way is at last brought down to the ground; and the greater the velocity is with which it is projected, the farther it goes before it falls to the earth. We may therefore suppose the velocity to be so increased that it would describe [follow] an arc [curve] of 1, 2, 5, 10, 100, 1000 miles before it arrived at

the earth, till at last, exceeding the limits of the earth, it should pass into space without touching it.

Let AFB represent the surface of the earth, C its centre, VD, VE, VF the curved lines which a body would describe, if projected in an horizontal direction from the top of an high mountain successively with more and more velocity . . . and . . . let us suppose . . . that there is no air about the earth . . . and for the same reason that the body projected with a less velocity describes the lesser arc VD, and with a greater velocity the greater arc VE, and augmenting [increasing] the velocity, it goes farther and farther to F and G, if the velocity was still more and more augmented, it would reach at last quite beyond the circumference of the earth, and return to the mountain from which it was projected . . . and . . . retaining the same velocity, it will describe the same curve over and over, by the same law . . . But if we now imagine bodies to be projected in the directions of lines parallel to the horizon from greater heights, as of 5, 10, 100, 1000 or more miles . . . those bodies, according to their different velocity, and the different force of gravity in different heights, will . . . go on revolving through the heavens in those orbits just as the planets do in their orbits.

DOCUMENT 8

EARTH'S POSITION IN THE UNIVERSE

NICOLAS COPERNICUS (1473-1543) – Trained in church law and medicine, he helped his uncle govern the small state of Ermland in East Europe, but he was far more interested in his hobby of astronomy.

That the Earth is not the centre of all revolutions is proved by the apparently irregular motions of the Planets and the variations in their distances from the Earth. These would be unintelligible if they moved in circles concentric with the Earth (that is with the Earth at their centre) . . . we shall place the Sun himself at the centre of the Universe. All this is suggested by the systematic procession of events and the harmony of the whole Universe, if only we face the facts, as they say, 'with both eyes open'. . . . We therefore assert that the centre of the Earth, carrying the Moon's path, passes in a great orbit among the other Planets in an annual revolution round the Sun; that near the Sun is the centre of the Universe; and that whereas the Sun is at rest, any apparent motion of the Sun can be better explained by the motion of the Earth. Yet so great is the Universe that though the distance of the Earth from the Sun is not insignificant . . . it is insignificant compared with the distance of the sphere of the fixed stars.

DOCUMENT 9

SCIENCE *FRANCIS BACON – Writing in his book* Novum Organum *(1620)*

Those who have handled sciences have been either men of experiment or men of dogmas. The men of experiment are like the ant: they only collect and use: the reasoners resemble spiders, who make cobwebs out of their own substance. But the bee takes a middle course; it gathers its material from the flowers of the garden and of the field, but transforms and digests it by a power of its own. Not unlike this is the true business of philosophy; for it neither relies solely or chiefly on the powers of the mind, nor does it take the matter which it gathers from natural history and mechanical experiments and lay it up in the memory whole, as it finds it, but lays it up in the understanding altered and digested. Therefore from a closer and purer league between these two faculties, the experimental and the rational (such as has never yet been made) much may be hoped.

DOCUMENT 10

DISCOURSE ON METHOD *RENE DESCARTES – French philosopher writing in 1637*

I was in Germany at the time; the fortune of war had called me there. The onset of the winter held me up in quarters in which I found no conversation to interest me. I spent the whole day shut up alone in a stove-heated room, and was at full liberty to discourse with myself about my own thoughts . . . .

I thought the following four (rules) would be enough, provided that I made a firm and constant resolution not to fail even once in the observance of them.

The first was never to accept anything as true if I had not evident knowledge of its being so.

The second, to divide each problem I examined into as many parts as was feasible, and as was requisite for its better solution.

The third, to direct my thoughts in an orderly way; beginning with the simplest objects, those most apt to be known, and ascending little by little, in steps as it were, to the knowledge of the most complex.

And the last, to make throughout such complete enumerations and such general surveys that I might be sure of leaving nothing out.

Those long chains of perfectly simple and easy reasonings by means of which geometers are accustomed to carry out their most difficult demonstrations had led me to fancy that everything that can fall under human knowledge forms a similar sequence.

DOCUMENT 11

INVESTIGATION BY EXPERIMENT *THOMAS SPRAT – Bishop of Rochester in the seventeenth century, who knew many of the scientists in the Royal Society, here describing their procedure in investigation.*

It has been their usual course, when they themselves appointed the Trial [the investigation] to propose one week, some particular Experiments, to be prosecuted [carried out] the next; and to debate before-hand, concerning all things that might conduce to the better carrying them on. In this Preliminary Collection, it has been the custom for any of the Society, to urge what came into their thoughts, or memories concerning them; either from the observations of others, or from Books, or from their Experience, or even from common Fame itself. And in performing this, they did not exercise any great rigour of choosing, and distinguishing between truths and falsehoods: but a mass altogether as they came; the certain Works, the Opinions, the Guesses, the Inventions . . . the Probabilities, the Problems, the general Conceptions, the miraculous Stories, the ordinary Productions . . . and whatever they found to have been begun, to have failed, to have succeeded, in the Matter which was then under their Disquisition [investigation].

DOCUMENT 12

INVESTIGATION BY MICROSCOPE *ANTONI VAN LEEUWENHOEK (1632-1723) – A Dutchman who made some of the most powerful microscopes of his day, here reporting a discovery in a letter published in the journal of the Royal Society*

In my letter of the 12th of September 1683, I spake, among other things, of the living creatures that are in the white matter which lieth . . . betwixt . . . one's front teeth or one's grinders. Since that time . . . I have examined this stuff divers times; but to my surprise, I could discern no living creatures in it.

Being unable to satisfy myself about this, I made up my mind to put my back into the job, and to look into the question as carefully as I could. But because I keep my teeth uncommon clean, rubbing them with salt every morning, and after meals generally picking them with a fowl's quill, or pen; I found very little of the stuff stuck on the outside of my front teeth: and in what I got out from between them, I could find nothing with life in it. Thereupon I took a little of the stuff that was on my . . . grinders; but though I had two or three shots at these observations, 'twas not till the third attempt that I saw one or two live animalcules . . . .

Having allowed my speculations to run on this subject for some time, methinks I have now got to the bottom of the dying-off of these animalcules. The reason is, I . . . pretty near always of a morning drink coffee . . . so hot that it puts me into a sweat: . . . Now the animalcules that are in the white matter on the teeth . . . being unable to bear the hotness of the coffee, are thereby killed: like I've often shown that the animalcules which are in water are made to die by a slight heating.

Accordingly, I took (with the help of a magnifying mirror) the stuff from off . . . the teeth further back in my mouth, where the heat of the coffee couldn't get at it. This stuff I mixt with a little spit out of my mouth . . . and then I saw with as great a wonderment as ever before, an unconceivably great number of little animalcules, and in so unbelievably small a quantity of the foresaid stuff, that those who didn't see it with their own eyes could scarce credit it. These animalcules, or most all of them, moved so nimbly among one another, that the whole stuff seemed alive and all moving.

DOCUMENT 13

THE ORIGIN OF THE SPECIES *CHARLES DARWIN – In an excerpt from his book (1859)*

When on board H.M.S. *Beagle,* as a naturalist, I was much struck with certain facts in the distribution of the organic beings inhabiting South America, and in the geological relations of the present to the past inhabitants of that continent. These facts . . . seemed to throw some light on the origin of species – that mystery of mysteries . . . . On my return home, it occurred to me, in 1837, that something might perhaps be made out on this question by patiently accumulating and reflecting on all sorts of facts which could possibly have any bearing on it . . . from that period to the present day I have steadily pursued the same object.

DOCUMENT 14

FIRST WORDS FROM THE MOON

NEIL ARMSTRONG – An American and the first man on the moon, 21st July, 1969

I'm going to step off the L.M. [Lunar Module] now . . . That's one small step for a man; one giant leap for mankind.

ACKNOWLEDGMENTS

Cambridge University Press pages 10 top, 15 bottom; Mary Evans Picture Library page 13 middle; H.M.S.O. page 3 top; *The Legacy of China,* Joseph Needham page 3 bottom; The Mansell Collection page 17; National Portrait Gallery pages 13 bottom, 16; Paul Popper Ltd page 13 top; *Punch* magazine page 20; Ronan Picture Library pages 10 top left, 11, 14 top left and top right, 15 top, 18; Science Museum pages 7, 8, 10 bottom right, 12, 14 bottom.

Greenhaven World History Program

History Makers
Alexander
Constantine
Leonardo Da Vinci
Columbus
Luther, Erasmus and Loyola
Napoleon
Bolivar
Adam Smith, Malthus and Marx
Darwin
Bismark
Henry Ford
Roosevelt
Stalin
Mao Tse-Tung
Gandhi
Nyerere and Nkrumah

Great Civilizations
The Ancient Near East
Ancient Greece
Pax Romana
The Middle Ages
Spices and Civilization
Chingis Khan and the Mongol Empire
Akbar and the Mughal Empire
Traditional China
Ancient America
Traditional Africa
Asoka and Indian Civilization
Mohammad and the Arab Empire
Ibin Sina and the Muslim World
Suleyman and the Ottoman Empire

Great Revolutions
The Neolithic Revolution
The Agricultural Revolution
The Scientific Revolution
The Industrial Revolution
The Communications Revolution
The American Revolution
The French Revolution
The Mexican Revolution
The Russian Revolution
The Chinese Revolution

Enduring Issues
Cities
Population
Health and Wealth
A World Economy
Law
Religion
Language
Education
The Family

Political and Social Movements
The Slave Trade
The Enlightenment
Imperialism
Nationalism
The British Raj and Indian Nationalism
The Growth of the State
The Suez Canal
The American Frontier
Japan's Modernization
Hitler's Reich
The Two World Wars
The Atom Bomb
The Cold War
The Wealth of Japan
Hollywood